ANIMALS REVIEWED

Starred Ratings
of Our Feathered, Finned,
and Furry Friends

Ratings contributed by the accredited facilities
of the Association of Zoos & Aquariums

ASSOCIATION
OF ZOOS &
AQUARIUMS

Timber Press ★ Portland, Oregon

Photo and illustration credits appear on page 178.

Published in 2019 by Timber Press, Inc.

The Haseltine Building
133 S.W. Second Avenue, Suite 450
Portland, Oregon 97204-3527
timberpress.com

Printed in China

Text and cover design by Kim Thwaits

Front cover photos, clockwise from upper left: Cincinnati Zoo / Teagan Dumont, Michael Durham / Oregon Zoo, Woodland Park Zoo, Woodland Park Zoo. Back cover photos, clockwise from upper left: Amber Paczkowski, Texas State Aquarium, Katie Canady, Amelia Beamish.

ISBN 978-1-60469-960-9

Catalog records for this book are available from the Library of Congress and the British Library.

INTRODUCTION

Is a sea otter worth it?

The answer is always yes. The same goes for pandas, pangolins, and penguins. Not to mention naked mole-rats, tardigrades, and lampreys. Because the truth is, every species is the G.O.A.T. at something.

It's hard to beat the pollinating power of the western bumblebee, the mosquito-zapping efficiency of the little myotis, or the unsavory yet critical cleanup job a turkey vulture performs.

But there's much more to these 5-star creatures than their ecosystem services. Hummingbirds, whale sharks, and the rest of their furred, feathered, and exoskeletoned kin make our planet a much more interesting place to live.

You can't put a price on a sea otter snuggling her fluffball pup, or the squeak-toy call of a pika.

But you can rate them.

The idea for #RateASpecies came to me while I was shopping online for hiking boots. The first review that caught my eye read "★★★★, waterproof, great for winter, only comes in brown." It was basically a description of my favorite Oregon Zoo resident, a rescued sea otter pup named Lincoln. From there, the tweets started writing themselves.

But it was the collective wit of the Association of Zoos and Aquariums (AZA) that set the hashtag aflame. Other conservation organizations and wildlife aficionados joined in to rate their favorite species, and the result was an outpouring of animal therapy from every corner of Twitter.

Every living thing on Earth has followed an epic evolutionary path to reach its current form: a high-definition masterpiece of nature. That they also evolved to bring us joy was a happy accident. The good news is that these durable, premium-quality species will continue to delight us for many years to come—if we care enough to make room for them.

Some of the species rated in this book, like the California condor, might no longer exist if it weren't for the collaborative conservation efforts of zoos and aquariums. In the United States and worldwide, AZA institutions are saving wildlife and the wild places we all depend on.

But our organization plays another, equally important role: inspiring the next generation of wildlife defenders. And I can't think of a better way to remind ourselves of what's at stake than a bunch of overzealous tweets of wildlife affection.

So thank you for reading and reviewing our world of wildlife, helping people everywhere make informed decisions about which species to fawn over next.

And feel free to join in and add your own contributions. We couldn't rate every species, but—in our book—every species rates.

Shervin Hess
Oregon Zoo

Linnaeus's two-toed sloth, *Choloepus didactylus*
TEXAS STATE AQUARIUM

★★★★★

It's furry, cute, and even comes with built-in hooks to easily hang it from nearly anything, but it quickly runs low on power and spends most of the day recharging. Still, totally worth it.

★★★★★

PRAYING THEY KEEP MAKING THESE

The ultimate seafood multitool: tenderizes, slices, and serves in milliseconds! Comes in all colors—plus some you can't see.

Peacock mantis shrimp, *Odontodactylus scyllarus*

MONTEREY BAY AQUARIUM

Cheetah, *Acinonyx jubatus*

DALLAS ZOO AND MARYLAND ZOO

★★★★★

FASTER THAN EXPECTED!

This thing tops out at highway speeds in seconds. And it's leaner and more sleek than pictured. Runs on just ground meat. Bonus: it purrs louder than my '85 Buick.

OCTO-MAZING!

This waterproof model expands to cover large spaces; customizable coloring blends easily with any décor. Eight arms equipped with suction cups secure to any surface. Bonus: super-collapsible for easy storage.

Giant Pacific octopus, *Enteroctopus dofleini*

ROSAMOND GIFFORD ZOO

Northern saw-whet owl,
Aegolius acadicus

OREGON ZOO

★★★★★

YOU WON'T BE DISAPPOINTED

This stylish little unit is amazing. Sound quality A+. No distortion at full volume, but bass is a little weak. Top rotates, which is a plus.

Its vibrant color and texture really pull our jungle look together, but minus two stars because of misleading common name—this green iguana is actually mostly orange-red. The scientific name is pretty uncreative, too. Iguana iguana? C'mon, guys.

Green iguana,
Iguana iguana

TEXAS STATE AQUARIUM

Red panda,
Ailurus fulgens

SEQUOIA PARK ZOO

★★★★★

DOMESTIC TRANQUILITY AT LAST

Every week it was the same complaint: "Dad, do I HAVE to prune the bamboo again? Just get a panda!" I'd explain that they're too big and expensive, and that it practically takes an act of Congress to get one. Then I found this compact model. What a great labor and time saver! I'm happy, the kids love its antics, and my yard is the best on the block, which doesn't go over well with certain neighbors. No, Mr. Johnson, I didn't "use hair dye on a darned raccoon."

PLEASED OVERALL

I ordered an audio recorder and received this. Downside, it only records what it feels like recording. Playback feature sometimes doesn't work, and there is no volume control. I decided to keep it because of the cool colors. Beware! Might peck at you if you try to fast forward.

Blue-and-gold macaw,
Ara ararauna

EL PASO ZOO

Tried ordering some new bongos for our weekly drum circle, and they sent us these giant deer? How do I even send these back? I CANNOT hold a beat on these things.

Eastern bongo,
Tragelaphus eurycerus isaaci

VIRGINIA ZOO AND NAPLES ZOO

African penguin, *Spheniscus demersus*

SENECA PARK ZOO AND RACINE ZOO

★☆☆☆☆

Ungrateful, fussy bird. Recently installed an igloo and industrial AC in my garage to make him feel at home—refused to go in and yelled like a donkey when I tried to take him outside for some fresh air on my snow day. Disappointed. Would not buy again.

★★★★☆

Top less "cottony" than expected and haircut a little outdated. Still, agree with other reviews—it's the cutest monkey in the world.

Cotton-top tamarin,
Saguinus oedipus

OAKLAND ZOO

Central bearded dragon,
Pogona vitticeps

JENKINSON'S AQUARIUM

With winter coming, I wanted a dragon that breathes fire! This one doesn't. However, it seems to be smiling at me and is content in a variety of habitats. Also eats what bugs him and enjoys a salad once in a while.

Giant anteater, *Myrmecophaga tridactyla*
GREENSBORO SCIENCE CENTER

★★★★★

Works well removing bugs from hard-to-reach places. If you're looking for an everyday vacuum, be forewarned: the nozzle does NOT detach. Leaves a slimy residue, but it does come with a bonus duster on the backside.

★☆☆☆☆

Too bad this product has been "unavailable" for YEARS! Bring it back, please! Its gold color is so beautiful, and I've also been told it's lucky. I'll keep waiting.

Panamanian golden frog, *Atelopus zeteki*

MARYLAND ZOO

★★★★☆

I WAS A DOUBTER

I was skeptical, but this really works! You can't put just any old kangaroo in a tree. You need to make sure you get the right kind. Look for labels like "Matschie's" or "Goodfellow's" to know it's the real deal.

Tree kangaroo,
Dendrolagus matschiei

WOODLAND PARK ZOO

Sandbar shark, *Carcharhinus plumbeus*

TEXAS STATE AQUARIUM

★★★★★

We're not going to leave a biting review here; this thing is totally jaw-some! It is pretty sharp on one end but still surprisingly safe. I can see why they've barely changed the design in 80 million years.

★★★★★

Know what's cute? A pig. Know what's cuter? A dainty pig with a spiky 'do. Sadly, it's on its way to becoming a collector's item. Maybe more high ratings will help spur production in its native land?

Chacoan peccary, *Catagonus wagneri*

SEQUOIA PARK ZOO

Arctic fox,
Vulpes lagopus

UTICA ZOO AND
MARYLAND ZOO

NOT A POMERANIAN

Have had some issues. Depending on what time of year it is, it changes color. It's impossible to spot during the winter, especially when snowy! Plus, it seems to either run out of power or rev up to and beyond ultra-high speed for no apparent reason.

★★★★★

PLEASED OVERALL! WOW!

Forget Dyson and Shark—this vacuum blows them all out of the water! Comes with long extendable nozzle with great reach! Sucks up food scraps without any hassle—but the process for emptying it out is a bit messy. Alarm feature is quite loud but easily adjusted by an ear rub.

Asian elephant, *Elephas maximus*

EL PASO ZOO

Lesser kudu, *Tragelaphus imberbis*

MARYLAND ZOO

★★★★☆

Try *greatest* kudu! The people who named this really undersold it. Very sleek, fast, and beautiful—just look at those accessories. Totally underrated.

GREAT VALUE

LOVE these. Must-have on all outdoor adventures. Keeps bugs off even when wet. Cordless.

Big brown bat,
Eptesicus fuscus

OREGON ZOO

Hippopotamus,
Hippopotamus amphibius

CINCINNATI ZOO

★★★★☆

BETTER THAN EXPECTED!

Arrived early and way below predicted weight but eventually reached normal size. Expected it to float and swim immediately, but it did neither. Subtracted one star for that. Is very photogenic and likes cameras. Fast learner and awesome ambassador. Comes in only one color, and shipping and maintenance is expensive.

★★★☆☆

TOO JUMPY

Skittish and up all night. Super cute, but came with an odd note: "Don't add water after midnight." Still not sure what this means. . . .

Mohol bushbaby, *Galago moholi*

ROSAMOND GIFFORD ZOO

★★★★☆

Stunning 2015 model. Larger than expected. A little moody in winter, but rack is great for holding hats, coats, scarves, etc. Keeps grass and shrubs trimmed nicely in the summer months but is a little rough on tree bark. The "bugle" setting is the perfect alarm call for any time of day. Excellent year-round product for a large family.

Wapiti,
Cervus canadensis

DAKOTA ZOO

Okapi,
Okapia johnstoni
POTAWATOMI ZOO AND DALLAS ZOO

Assembly instructions sorely lacking, but I *think* it turned out okay. Had to use parts from other projects to complete—fortunately I had spare ossicones and legs. Ears feel like velvet, and the 14-inch tongue reaches top pantry shelf with no problem. Delighted to finally have one of these special models.

Not very hog-like. . . . Almost no snout to speak of, very spiky, and super small. Pretty adorable, likes my hedgerows, and emits little piggy grunts, though!

Four-toed hedgehog,
Atelerix albiventris

CHEYENNE MOUNTAIN ZOO

Toco toucan,
Ramphastos toco

RIVERBANKS ZOO & GARDEN

★☆☆☆☆

FALSE ADVERTISING

I was totally misled. . . . I thought it would like fruity cereal. . . . Prefers all-natural fruit instead. Do yourself a favor and order a good pair of sunglasses—those colors are even brighter in person!

★★★★★

GREAT BOGO!

Pleasantly surprised to receive two fuel tanks for the price of one. Runs well on biofuel (hay, grain, carrots, yams). 2017 model has all-weather coverage and adaptability. Relatively easy to operate, though temperamental at times.

Bactrian camel,
Camelus bactrianus

ROSAMOND GIFFORD ZOO
AND POTAWATOMI ZOO

Llama,
Lama glama

SEQUOIA PARK ZOO

Seeking the wisdom of a spiritual leader, I ordered myself a llama. It was a lot hairier than I expected and looked more like a humpless camel than a wise Tibetan. Whenever I ask it a question, it looks deep in my eyes and its lips move, yet it says nothing. Stuck on mute?

With its aerodynamic shape, built-in sonar, and ability to reach speeds of up to 20 mph, this *fin*-tastic model is basically a submarine powered by fish. What more could you ask for? Expensive to refuel, though, so expect to have to fork over as much as 20 pounds of herring a day.

Atlantic bottlenose dolphin, *Tursiops truncatus*

TEXAS STATE AQUARIUM

I wanted something to protect my property, but this lazy pair of black-and-white slobs ain't the answer. They sleep all day, will eat you out of house and home, get into everything, and don't even spray unless you really, really scare them! Even then, they are more likely to run away than stand their ground! I'm trying a guard rabbit next time—it would do a better job!

Striped skunk,
Mephitis mephitis

CHEYENNE
MOUNTAIN ZOO

Giant Flemish rabbit,
Oryctolagus cuniculus

MARYLAND ZOO

★★★☆☆

Runs large! Check the sizing chart before ordering. I'm usually a medium but ordered giant because it was the only size in stock. It's huge! Still gave three stars because the shape is great and it's *sooo* soft.

★☆☆☆☆

Description said it would eat rodents, so I bought some for my Florida backyard rat problem years ago. They immediately got out of the fence. Very disappointed and would not recommend for outdoor use.

Burmese python,
Python bivittatus

ZOO MIAMI

Eastern black rhinoceros, *Diceros bicornis michaeli*

BLANK PARK ZOO

★★★★★

The most intense CrossFit instructor ever! Unusual approach to exercise—I could barely keep up. Definitely will be back for more!

★★★★☆

Not what I expected, but fun! Advertised as "retro-style guinea pig," but I just don't see it. More like a short-eared jackrabbit on steroids, only cuter. No need to buy more than one pair—they'll make all you need.

Patagonian cavy, *Dolichotis patagonum*

SEQUOIA PARK ZOO

★★★☆☆

There must have been some kind of shipping accident—product arrived flattened, with mouth and eyes on separate sides. Still, we love the winning smile, and it's not nearly as dangerous as other reviews might lead you to believe.

Cownose ray,
Rhinoptera bonasus

TEXAS STATE AQUARIUM

Andean condor, *Vultur gryphus*

TRACY AVIARY

★★★★★

Highly recommended for even the worst biohazard cleanup—this white-ruffed guy goes gangbusters. He'll chew up any disease-laden remains you throw at him, and—surprise, surprise—he's self-cleaning. Have had ours for more than 60 years and he's just as sturdy as day one.

★★★★★

Super happy with this guy. Slightly lazy at times, motivated by food, but has NEVER skipped leg day.

Komodo dragon, *Varanus komodoensis*

VIRGINIA AQUARIUM & MARINE SCIENCE CENTER AND ZOO ATLANTA

Southern sea otter,
Enhydra lutris nereis
OREGON ZOO

★★★★★

FIRST IMPRESSIONS

Overall very good. Sturdily built, totally winter-ready and waterproof. Only comes in brown but that's actually a plus for me.

★★★☆☆

MISSING PARTS

When assembling, found that none of them had thumbs. Sent back. The next ones had the same issue. Had to call manufacturer. Apparently, this model doesn't have them at all. Need to be clearer on product specification! Chose to keep because they're pretty quiet and visually appealing.

Eastern black-and-white colobus, *Colobus guereza*

CHEYENNE MOUNTAIN ZOO AND OREGON ZOO

★★★☆☆

NEEDS INSTRUCTIONS

I've long admired my neighbor's flamingos so decided to get some of my own. But I can't figure out how to get them to stay still. His flock is quiet and never seems to move, while mine are a hub of nonstop activity. I worry about their stability, but they balance well even on one leg, and the color still hasn't faded despite their spending most of the time in water.

Chilean flamingo,
Phoenicopterus chilensis

SEQUOIA PARK ZOO AND GREAT PLAINS ZOO & DELBRIDGE MUSEUM OF NATURAL HISTORY

Black bear,
Ursus americanus

OAKLAND ZOO

★★★★☆

Name misleading—not dark black as marketed. If you ever think you've misplaced it, look up! And be careful not to leave any leftovers out. Overall great value for size.

Perfect aquarium fish . . . except guests keep saying they found Dory. Not sure why because this one's name is Debra.

Regal blue tang,
Paracanthurus hepatus

RIPLEY'S AQUARIUM OF MYRTLE BEACH AND FORT WAYNE CHILDREN'S ZOO

American alligator, *Alligator mississippiensis*

TEXAS STATE AQUARIUM

★★★★☆

OH SNAP!

This model is long-lasting with a really tough exterior, but unfortunately I've lost it several times—its color easily blends in with its surroundings. Take note, this is different from the similar-looking "crocodile."

★★★★☆

GREAT DANCER

Dances well during courtship. Poor eyesight, but tiny sensory hairs make up for that.

Emperor scorpion,
Pandinus imperator

JOHN BALL ZOO

There was some difficulty getting it through the front door, otherwise perfect in every way.

Cuteness: 10/10
Social skills: 10/10
Durability: 10/10
Ability to be very tall: 10/10
Uniqueness: 10/10
Limbo competitions: 0/10 (But seriously, who even likes to do the limbo?)

Giraffe,
Giraffa tippelskirchi

NORTH CAROLINA ZOO
AND PHOENIX ZOO

Green sea turtle,
Chelonia mydas

TEXAS STATE AQUARIUM AND SOUTH CAROLINA AQUARIUM

★★★★★

TAKES AMAZING "SHELLFIES"

Lightweight, ergonomic design, waterproof. Comes with a beautiful protective patterned case, and can last as long as 100 years!

AS DESCRIBED

Bought these out of curiosity. Arrived clumped together. Fur was missing. Blind as bats and sleep all day. I could have deducted another star, but these things work hard at night, running forward and backward and building tunnels. Available only in nude, making them seriously #NSFW.

Naked mole-rat, *Heterocephalus glaber*

ROSAMOND GIFFORD ZOO, SENECA PARK ZOO, AND RIVERBANKS ZOO & GARDEN

Baird's tapir,
Tapirus bairdii

PALM BEACH ZOO

★★★★★

WORTH THE WAIT!

New models come with spots and stripes, and they look fantastic! The manufacturer only makes one every 400 days or so, so we are lucky to have it. Would highly recommend!

A TOUR DE FORCE

Simply superlative. Disruptive design. Inspired mechanical engineering. Invented its own niche with seamless software/hardware integration. Absolute head-turner and conversation starter. Sorry, no headphone jack: the future is here.

Ocean sunfish,
Mola mola

MONTEREY BAY AQUARIUM

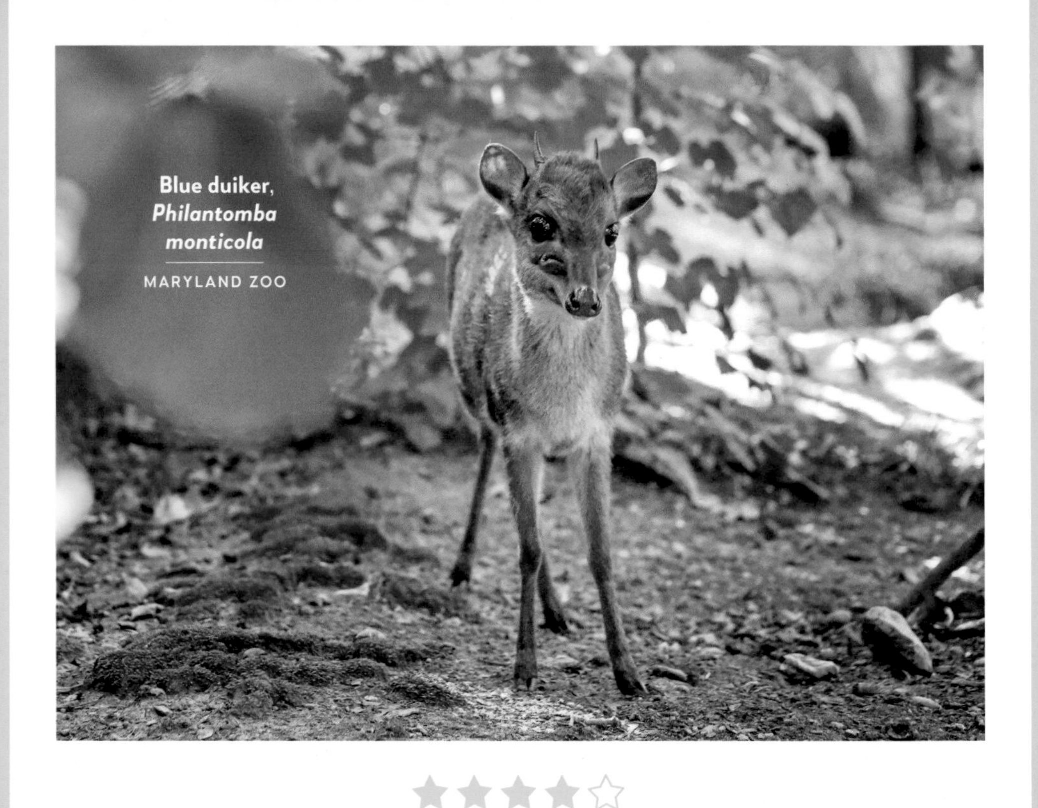

Blue duiker,
Philantomba monticola

MARYLAND ZOO

★★★★☆

As a miniatures enthusiast, I consider this a top-notch addition to my collection. Very small size, excellent detail in the tiny horns and teeny hooves. I'm pleased.

★☆☆☆☆

Looks cute, but size doesn't match description. Buyer beware that this item is NOT giant! Says it runs on bamboo, but no matter how much I feed it, it always ends up in sleep mode. Also has so much personality I'm still not convinced it's not just a person in a bear suit.

Giant panda, *Ailuropoda melanoleuca*

ZOO ATLANTA

American beaver,
Castor canadensis

OREGON ZOO

★★★★★

SO MUCH FUN

Cuts through standing water like butter. Perfect for lakes and ponds. Setup is easy, and it won't flip. Variable speed. No mount for GoPro.

★★★★☆

VERY DURABLE

I even dropped this thing and it still works! It's built like a tank—military grade for sure. I will probably upgrade to the 6-banded version soon.

Southern three-banded armadillo, *Tolypeutes matacus*

DALLAS ZOO

Eastern loggerhead shrike, ***Lanius ludovicianus migrans***

TORONTO ZOO

★★★★★

Impressive range of functionality, plus surprisingly good sound quality.

- Unconventional, but very effective pest control! (Not compatible with mouse.)
- May contain scenes of violence.
- Getting harder to find, but nice to have for the summer. I would consider a different option for winter.
- You will never have to buy another hole punch!

★★☆☆☆

Beautiful packaging, but scent is off-putting. Notes of fermented cabbage and sulphur with undertones of onion.

Maned wolf, *Chrysocyon brachyurus*

DICKERSON PARK ZOO

Orangutan,
Pongo

RACINE ZOO

★★★★★

Wonderful product! Very strong. Was lucky enough to get one before they run out! Apparently the place that makes them is disappearing.

Get the "steller" edition. This limited-edition series is the F-150 of pinnipeds. Buyer beware: premium fuel costs will burn a hole in your wallet, but the performance is unmatched. Removing 1 star as ours has started to smell strongly. No muffler.

Steller sea lion,
Eumetopias jubatus

VANCOUVER AQUARIUM

Tawny frogmouth,
Podargus strigoides

TRACY AVIARY

★★★★★

BEST KNOCK OFF ON THE MARKET!

Most people will never realize you didn't spend the money on a name-brand owl. Besides, much softer to hold than the real thing. Still works well with the insects, but had to throw out most of the mice it came with.

★★★★☆

TRUSTED NAME IN 24/7 PROTECTION

Tough on bad guys. With over 30,000 quills to inflict pain, it's truly cuteness in disguise. Does not come bubble-wrapped, so HANDLE WITH CARE! (Docked a star for no safety label.)

African crested porcupine, *Hystrix cristata*

OKLAHOMA CITY ZOO AND DOWNTOWN AQUARIUM DENVER

Uracoan rattlesnake, *Crotalus durissus vegrandis*

EL PASO ZOO

Not pleased. Assumed would be friendly because of wagging tail. NOT FRIENDLY AT ALL. One star because it effectively controls mice population.

This little crab is entertaining, but I'm giving it only two stars because apparently its fiddle is an accessory that is sold separately. My poor little crab still tries to pantomime playing without it. Even waves its oversized claw at me sometimes to encourage me to watch.

Atlantic marsh fiddler crab, *Uca pugnax*

MARITIME AQUARIUM

★★★★★

POUNCE ON THIS ONE

Ad stated, “small and furry with a wonderful purr.” Started out small but got stretched out over time. And the purr is more like RRRRRRRRRRRRRRRRRRRRRRRRR. Still, even after heavy use in all weather, very tough, very hardy.

Snow leopard,
Panthera uncia

SENECA PARK ZOO AND WOODLAND PARK ZOO

Galapagos tortoise, *Chelonoidis nigra vaudenburghi*

EL PASO ZOO AND RIVERBANKS ZOO & GARDEN

★★★★★

TALK ABOUT AN INVESTMENT

FAIR WARNING—You'll think they look quite compact when they arrive, but these tortoises grow to be MASSIVE. Come with a heavy-duty case that'll last for over a century. Manufactured exclusively on a few remote islands off Ecuador.

★☆☆☆☆

WAY TOO SMALL

Very disappointed in the size! Apparently "lesser" means MUCH less—which they must know at the factory because they tried to throw in a mini extra version to make up for it. Also underestimated how bad it would smell. Traded it in for the giant version, only had to pay for shipping

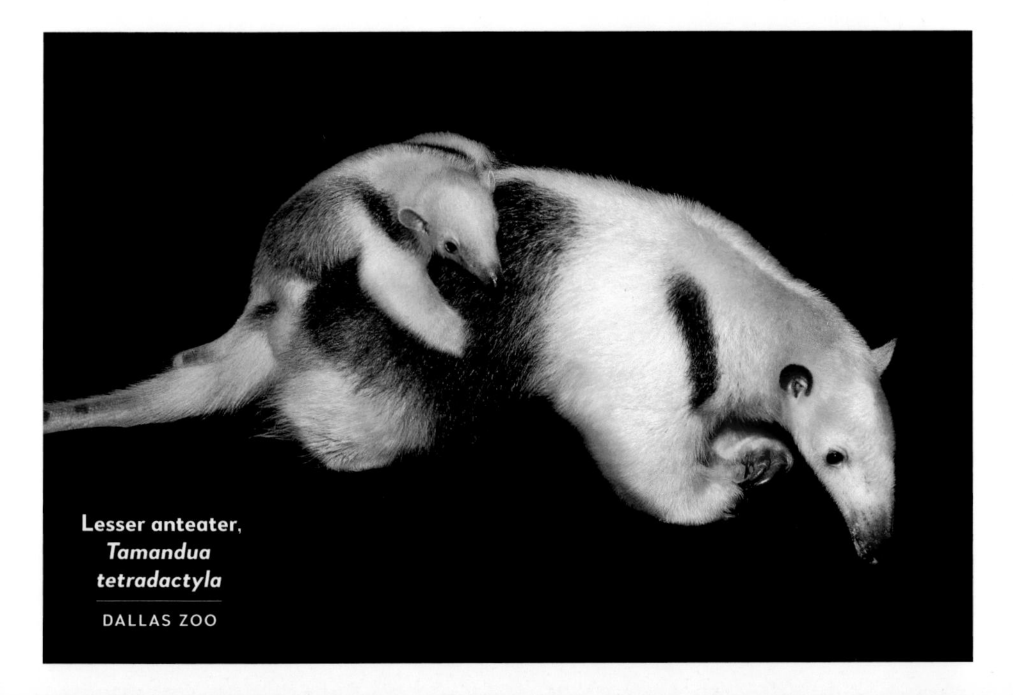

Lesser anteater, *Tamandua tetradactyla*

DALLAS ZOO

Scimitar-horned oryx, *Oryx dammah*
LEHIGH VALLEY ZOO

★★★★★

EXTREMELY RARE FIND!

Very adaptable with a built-in thermostat to withstand warm and cool temperatures. Beautiful long horns but prefers to run away, so not really the best security system. Has adorable offspring—added a star for cuteness.

NOT AS EXPECTED

Meant to order new serving spoons for my upcoming party but got something way different. Great for stirring drinks, eating ice cream, and scooping dips. Unfortunately only comes in one model. Waiting for fork and knife versions to become available.

African spoonbill, *Platalea alba*

OAKLAND ZOO AND OREGON ZOO

Excellent product. Dug the holes I needed (and some I didn't realize I needed) in an hour and a half. Bonus: haven't seen a bug in days. Highly recommended.

Slender-tailed meerkat, *Suricata suricatta*

CHEYENNE MOUNTAIN ZOO

Largescale four-eyed fish,
Anableps anableps

ZOO MIAMI

Advertised as four-eyed fish. Came with only TWO eyes! Should be called four-pupiled fish. Keeps climbing out of water onto rocks in aquarium LIKE A FISH TOTALLY SHOULDN'T!!

GRAZES LIKE A BOS!

Anybody can have a cow, and that's just fine. But if you wanna be the coolest zoo on the block, you need the biggest, baddest, shaggiest bovine around—the yak! Everyone will be jealous of your Himalayan Holstein.

Yak,
Bos grunniens

SEQUOIA PARK ZOO AND DETROIT ZOO

Pacific sea nettle,
Chrysaora fuscescens

TEXAS STATE AQUARIUM

This version doesn't seem to come with any kind of steering. Still, it's strangely fascinating to watch—I'll take this over a lava lamp any day. Warning: do not touch these with bare hands. Trust me.

White-handed gibbon, *Hylobates lar*

RACINE ZOO

★★★★★

Excellent product, has exceeded my expectations! Instructions said it would need to be replaced at least every 30 years, but mine is now over 50 and still working spectacularly!

I ordered the rare salamander starter kit and everything looked good at first, but they just stayed tadpoles. Now I am stuck with these giant tadpoles that grew legs but never moved to land.

Axolotl,
Ambystoma mexicanum

CHEYENNE MOUNTAIN ZOO AND DALLAS WORLD AQUARIUM

Hoffmann's two-toed sloth, *Choloepus hoffmanni*

ZOO ATLANTA

★★★★☆

FAULTY MODEL?

New arrival appears to have three toes instead of two? However, it does have two fingers. Will probably keep because it doesn't make too much noise and is fun to hang out with.

★★★★★

WHAT A DEAL

Cleanup is a breeze with one of these. Gets rid of really strong odors. WAY bigger than I expected. So glad they didn't stop making them. On backorder now.

California condor, *Gymnogyps californianus*

OREGON ZOO

Largetooth sawfish,
Pristis pristis

RIPLEY'S AQUARIUM
OF MYRTLE BEACH

★★★☆☆

Band saw on the blink? Let a sawfish take a whack at your next woodworking project. The name is misleading, though. He's not actually the best at sawing. Also suggest better packaging—the box arrived shredded.

Thought I ordered a solid black bear, but this one came with a brown face. Called the manufacturer, and they tried to pass this off as "spectacles."

Andean (spectacled) bear,
Tremarctos ornatus

RACINE ZOO

White rhinoceros, *Ceratotherium simum*
LION COUNTRY SAFARI

★★★☆☆

NOT AN ACTUAL UNICORN

Always wanted to see a real, live unicorn. Order arrived 16 months later, no horn, and in gray. Likes mudbaths, keeps the lawn cut, and squeals a lot. Chubbier than expected . . . pretty cool though!

Ordered this so I'd have someone to watch the Arizona Diamondbacks games with me. This handsome turtle does not come with the team logo on it—or any sort of team paraphernalia—but it does seem to enjoy the D-Backs, particularly with a plate of fish and snail nachos.

Diamondback terrapin, *Malaclemys terrapin*

MARITIME AQUARIUM

Red-necked wallaby,
Macropus rufogriseus

CHEYENNE MOUNTAIN ZOO

★★☆☆☆

Ordered one of these to be my new best friend. Nine months later, a smaller, balder version started poking its head out of the first one's stomach! Has it been hiding in there, listening to our conversations, this entire time?! Now I have two to take care of. Plus, pretty sure that's an invasion of privacy.

ALWAYS ON SAIL

If you're an Enya fan or a beach bum, you need these ASAP. Left- and right-handed editions, depending on location. You can definitely afford to pick one up—nowhere near as painful as larger models. Pictured here: blue global edition.

By-the-wind sailor, *Velella velella*

MONTEREY BAY AQUARIUM

African lion,
Panthera leo

UTICA ZOO

★★★★★

Recently bought this multipack and have been very happy with my purchase. Two of them look similar, but they are definitely unique. I ended up needing more space in my house for them as they were much bigger than anticipated. Be careful of sharp edges, and cover your ears around mealtime—they can get very loud!

Moluccan cockatoo, *Cacatua moluccensis*

JENKINSON'S AQUARIUM

Although it's not the home alarm system that I expected, it alerts me to absolutely everything with a sound comparable to a jet engine! Warranty indicates this model could last over 60 years. Sharp parts included—requires supervision when patrolling and should not be left alone with children or pets.

★★★☆☆

Description promised "up to 220 pounds"; arrived about the size of a tennis ball. Description also promised up to 70 years with limited warranty; only time will tell. Package did include an attractive pink lure, so there's potential for fish population control. Limited quantities. Production currently on the rise with increased distribution in Illinois.

Alligator snapping turtle, ***Macroclemys temminckii***

PEORIA ZOO

American white pelican, *Pelecanus erythrorhynchos*

JOHN BALL ZOO

★★★★☆

SPACIOUS

Comes in half a dozen varieties, and the lower half of its bill can hold up to 3 gallons of water! The pouch doesn't actually store food, though—minus one star.

★★★★★

HIGHLY RECOMMEND!

Waterproof and durable. Perfect for the beach but also works well on land. Compatible with older models and new editions. We just can't get enough!

California sea lion,
Zalophus californianus

SENECA PARK ZOO

Fossa, *Cryptoprocta ferox*

NAPLES ZOO

★★★☆☆

THOUGHT IT WAS A CAT... IT'S NOT!

Incredible balance and strength. But will climb on everything!! Also, advertised as "largest predator on Madagascar," but isn't much larger than a house cat! And the tail reminds me of a monkey?? Rated three stars for being unique. Took off two stars for misleading ad.

INCORRECTLY LABELED

Didn't know a single knock-knock joke! Not funny at all.

Clown anemonefish,
Amphiprion ocellaris

SEATTLE AQUARIUM

Bush dog,
Speothos venaticus

SEQUOIA PARK ZOO

INACCURATE PRODUCT DESCRIPTION

Ordered a dog, what we got was some sort of weird bear/weasel hybrid. Squeaks instead of barking. And what's with the smell? Did someone spill salad dressing?

UPDATE: ★★★★★

Active, adorable, and visitors are fascinated, so we will be keeping.

Laughing kookaburra, *Dacelo novaeguineae*

MARYLAND ZOO AND ZOO ATLANTA

★★★☆☆

NO SENSE OF HUMOR!

Have had for several days, has not laughed at any of my jokes. Maybe it's just sick and needs some tweetment. Louder than expected.

★★★★☆

Great color and texture. Very tree-like! Its super-adhesive feet make up for the fact that it is not anywhere near as large as its name falsely advertises. Still, would recommend.

Bavay's giant gecko, *Mniarogekko chahoua*

CHEYENNE MOUNTAIN ZOO

Bobcat,
Lynx rufus
MARYLAND ZOO

TOTALLY MISLEADING NAME!

I have been calling to "Bob" for days now with a 0% success rate. Plus, "Bob" sleeps all day. A disappointment.

VERY VERSATILE

This model enlarges easily and quickly with little effort from user. Minus one star because the deflating process takes wayyyyy too long. Not safe for kids!

Porcupine balloonfish, *Diodon holocanthus*

SEATTLE AQUARIUM AND OREGON COAST AQUARIUM

Brazilian tapir,
Tapirus terrestris

LION COUNTRY SAFARI

QUALITY DOESN'T TAPIR OFF

Very happy with our rhino-pig. Pleasant disposition. Will eat most fruits and vegetables. Enjoys bathing and has decent hygiene. Minus one star because ours doesn't oink.

★★☆☆☆

OBVIOUS SCAM

I thought flamingos were supposed to be tall and pink. Instead, I received a white fluffball with short little legs. Looks more like a gosling. Fake flamingo!

Chilean flamingo, *Phoenicopterus chilensis*

ZOO ATLANTA

Chuckwalla, *Sauromalus ater*

CALIFORNIA SCIENCE CENTER

★★★★★

Ordered a chuckwalla, received a Cirque du Soleil acrobat. When this showboating lizard isn't performing for audiences, he's doing pushups to assert his dominance. 10/10 would drive to Vegas to see this chuck's show.

★★★★☆

If you're like me and only buy American, you'll love this product. Some are made in Canada and northern Mexico, so be sure to ask your seller for a certificate of authenticity!

Bald eagle, *Haliaeetus leucocephalus*

MARYLAND ZOO

Black oystercatcher,
Haematopus bachmani

SEATTLE AQUARIUM

★★☆☆☆

FALSE ADVERTISING!

Prefers mussels. Very disappointing—will need to purchase a second product for catching oysters, which are growing uncontrollably all over my beach.

★★★☆☆

NOT AS DESCRIBED

Still not sure what I purchased—resembles a combination of a goat and a moose? The ears appear to be installed upside down. Climbed up on the roof and won't come down.

Sichuan takin, *Budorcas taxicolor tibetana*

ROSAMOND GIFFORD ZOO

★★★★★

WHAT A FORTUITOUS MISTAKE!

I was craving a sandwich, so searched for "natural ham" and hit one-click ordering before reading the full product description. Oops! It's true that these funny guys live to perform for an audience. No, they won't float on their back pounding open shellfish, but they will swim around your children in a tube. Take that, sea otter!

North American river otter, *Lontra canadensis*

SEQUOIA PARK ZOO

Arabian camel, *Camelus dromedarius*

JOHN BALL ZOO

★★★★☆

A LITTLE LUMPY AT FIRST

Minus one star for discomfort, but can go a long time without food or water—bonus.

NOTHING ELSE MATTERS

Get back to basics, people! These things provide kids with HOURS of entertainment wherever we go. They have dozens of models for free at our local aquarium. Forget sharks and penguins—these are what the kids really want.

Rock pigeon, *Columba livia*

MONTEREY BAY AQUARIUM

Zebra shark,
Stegostoma fasciatum

RIPLEY'S AQUARIUM OF CANADA AND FORT WAYNE CHILDREN'S ZOO

★★★★☆

Ordered a zebra shark but was sent this. Maybe he hasn't earned his stripes?

Camouflages easily off-road—but it sticks out like a sore thumb on New Jersey asphalt. If you decide to pick it up and help it across the street, be sure to put it on the side where it was headed!

Eastern box turtle,
Terrapene carolina

JENKINSON'S AQUARIUM

Virginia opossum, *Didelphis virginiana*

SEQUOIA PARK ZOO AND TENNESSEE AQUARIUM

★★★★★

If you're like me, you've been burned by other models that are short on teeth. Not this one! They packed more teeth into this than any similar product, 50 of them! Includes ingenious pocket for holding small items. Very cute marsupial—and the only one you can find in North America! Do negative reviewers know about the benefits to the ecosystem?

★★★★★

Bought him for my teenage son since they speak the same language. Perfect.

Grunt,
Haemulon flavolineatum

MOTE MARINE LABORATORY & AQUARIUM

I bought this new lawnmower to replace my previous machine. Does a great job of cutting the grass! Thought the fertilizer-application feature was a nice addition. . . until I stepped in some.

Nubian goat,
Capra hircus

COSLEY ZOO

Golden lion tamarin, *Leontopithecus rosalia*

CHEYENNE MOUNTAIN ZOO

★★☆☆☆

Incorrect audio. While the color, mane, and teeth are spectacular and very lion-like, product is much smaller than expected. Instead of roaring nobly, emits a high-pitched scream.

★★★★★

Ever had a pesky pickle jar that you just can't seem to open? This jar opener was the perfect solution for me! Extremely compact storage as well—can squish to the size of a quarter! Color- and texture-changing feature is an excellent bonus.

California two-spot octopus, ***Octopus bimaculatus***

CALIFORNIA SCIENCE CENTER

Kea,
Nestor notabilis

TRACY AVIARY

★★★☆☆

CAME WITHOUT INSTRUCTIONS

Can't seem to turn off the licking feature. We also haven't slept for several nights. It plays the same repetitive clanking song when given any musical toys and laughs when we get close. However, it makes us happy, so we'd reorder.

★☆☆☆☆

PLEASE ADD DIMENSIONS TO THE PRODUCT!

Wanted full-size kangaroo around 5 feet tall. Got something half that size.

Tammar wallaby, ***Macropus eugenii***

DAKOTA ZOO

Crocodile monitor, *Varanus salvadorii*

CENTRAL FLORIDA ZOO

This "security system" is a major fail! It doesn't even look like a crocodile so it's not scary, and it doesn't "monitor" the premises at all. Instead of being aquatic, he's always up in our trees, and his tail is always wrapping around things. I'd give him one star for functionality, but he IS pretty cute so adding a second.

★★★★☆

Pros: Extremely boopable snoot. Great climber. Makes very cute sounds when eating.
Cons: Not a good cuddle buddy. Smells like bad body odor. Requires additional equipment—gloves!

North American porcupine, *Erethizon dorsatum*

SEQUOIA PARK ZOO

Humboldt penguin,
Spheniscus humboldti

OREGON ZOO

★★★★★

THOUGHT IT WAS A KNOCKOFF

Looked like this right out of the package.
Started beeping, but they said that was normal.
Craftsmanship seems good. Fun.

★★★★☆

MORE THAN MEETS THE EYE!

Flashy and powerful. Transforms from female to male—but not into a crime-fighting robot.

Bumblebee grouper,
Epinephelus lanceolatus

MYSTIC AQUARIUM

Moon jelly,
Aurelia aurita

SEATTLE AQUARIUM

STRANGE FLAVOR but pretty to look at. Would prefer grape or strawberry. Can't figure out how anyone would spread this on toast.

★★★★☆

I guess it's partly my fault, but nothing worse than getting a brand new kite lost in the trees immediately!

Paper kite,
Idea leuconoe

SOPHIA M. SACHS BUTTERFLY HOUSE

Spiny dogfish, *Squalus acanthias*

SEATTLE AQUARIUM

★★★★☆

Minus one star because he won't play fetch.

Seems uninterested in fish. Definitely not cuddly. Keeps trying to eat my pet bunny, and wakes up suuuuuper early and goes to bed really late. Would return but it's currently chasing a porcupine, for some reason . . . I'll stay away for the moment!

Fisher,
Martes pennanti

SQUAM LAKES NATURAL SCIENCE CENTER

Leopard gecko,
Eublepharis macularius

RACINE ZOO

★★★★☆

Other reviewers have complained that the tail breaks off this product if it gets scared, but don't worry, it grows back!

Sounded cuddly, so I ordered one. He arrived and is SLIMY, which doesn't make for a great snuggle. We do enjoy a nice shrimp cocktail together though.

Dwarf cuttlefish, *Sepia bandensis*

MOTE MARINE LABORATORY & AQUARIUM

Glass lizard, *Ophisaurus*

CHEYENNE MOUNTAIN ZOO

NOT WHAT I ORDERED!

I wanted a lizard but got a snake instead. When I complained to the company, they claimed they sent the right product because of the eyelids and ears. What terrible customer service! This thing doesn't even have legs!

★★★★☆

USE AT NIGHT

Thought it was broken at first because it doesn't move much during the day. Almost returned, but noticed a change after dark. Twirling head movement was a cool surprise. Could work on its facial expressions, though.

Great horned owl, *Bubo virginianus*

TRACY AVIARY

Scarlet macaw, *Ara macao*

PHOENIX ZOO

★★★☆☆

Instantly mesmerized by the brilliant feathers, but was startled when the bird started talking back!

It dwells in isolated streams among rugged mountains. It awakens at night to feed, sometimes to consume one of its own kind. The name strikes fear in your heart. But . . . it looks like a cross between a mud puddle and a lasagna noodle. Named mine Jake. He's pretty chill.

Hellbender, *Cryptobranchus alleganiensis*

SENECA PARK ZOO AND TENNESSEE AQUARIUM

Sand tiger shark,
Carcharias taurus

JENKINSON'S AQUARIUM

The "shark" I was expecting was for floor cleaning. The teeth are a little overwhelming at first, and suction is somewhat lacking on hardwood floors, but this very laid-back version has done a wonderful job keeping our oceans healthy! Availability may be limited in future due to a demand for its fins.

Couldn't find any nearby, even though I hear of people seeing them all the time. So I ordered this one. Came all the way from Montana. Great condition, but bigger than expected. Weird meows, sounds like whistles and chirps—hope it's not broken.

Mountain lion,
Puma concolor

SQUAM LAKES NATURAL SCIENCE CENTER

Cockatiel,
Nymphicus hollandicus

CHEYENNE MOUNTAIN ZOO

Sound quality is good, but the power and sound buttons appear to be broken! Product always operates at top volume despite efforts to adjust audio.

★★★★★

VERY EFFICIENT

Filters nutrients from sand exactly as advertised. No attachments required. One port for all functions!

California sea cucumber, *Parastichopus californicus*

SEATTLE AQUARIUM

Southern screamer, *Chauna torquata*

ZOO MIAMI

★★★★☆

Asked Alexa to order chajá, an Uruguayan cake. Instead this showed up—apparently has the same name? Screams at everything that moves and came with large intimidating spurs on its wings. I'm still hungry, but now I have an amazing home security system!

★★★★★

CAN'T BEAT A CLASSIC!

Checks all the boxes for real freshwater-fish lovers. Very impressive that it hasn't needed updates since the time of the dinosaurs!

Lake sturgeon,
Acipenser fulvescens

TENNESSEE AQUARIUM

★★★★★

Ordered pink-necked model.
Received surprise sorbet variety pack.
Very satisfied!

Pink-necked green pigeon,
Treron vernans

VIRGINIA ZOO

★★★★☆

Product as advertised: a fish that sits around like a lump on a log. Named mine after my ex because its clever pelvic-fin adaptation creates a strong suction that keeps the fish securely in place—just like beer and TV did for Harold.

Lumpfish, *Cyclopterus lumpus*

MARITIME AQUARIUM

Fiji banded iguana, *Brachylophus bulabula*

VIRGINIA ZOO

★★★★★

Awesome lizard! Does anyone know if they need to drink only Fiji Water, or is tap water OK? Instructions didn't say.

Totally nails the Blue Steel look, but still working to refine Magnum. Likes selfies, mirrors, and gazing into her own eyes. In her defense, she's a total fox.

Red fox,
Vulpes vulpes

GREAT PLAINS ZOO & DELBRIDGE MUSEUM OF NATURAL HISTORY

★★★★☆

Forget the Roomba. This will revamp your spring cleaning. 1 star removed as it's aesthetically lackluster and carcass-powered. While not always practical, it can clear a whale-size mess, no problem.

★★★☆☆

Seems to be some kind of a mix-up from the shipping company? We received one that was labeled male . . . but now it's pregnant? I like that the prehensile tail means you can attach it to anything. With only a few small fins, though, it is far too slow to say it has any "horse" power!

Lined seahorse, *Hippocampus erectus*

TEXAS STATE AQUARIUM

★★★★★

SUPERSIZE ME!

This industrial-size model puts smaller RC aircraft to shame. Sturdy, fast, with a 5-foot wingspan and high-grip landing gear, this beauty flies with zero engine noise. Head rotates 270 degrees in either direction for a panoramic view.

Eurasian eagle-owl, ***Bubo bubo***

ROSAMOND GIFFORD ZOO

Giant day gecko, *Phelsuma grandis*

SENECA PARK ZOO

★☆☆☆☆

MISLEADING!

Bought this as a stand-in for the dragon in my next D&D game. Total waste of money. It's like a sticky, green, foot-long hotdog with judging eyes—not a majestic lizard king. Camouflage is cool, but we lost it in Kyle's mom's ficus.

★★★★☆

UNDERRATED

Don't let the strange packaging deter you—the disposal abilities here are superb. The bald head gets hard-to-reach places, leaving no trace of rabies or anthrax. Built-in self-cooling feet mechanism prevents overheating. Wanted the largest U.S. model but they're pretty hard to come by.

Turkey vulture, *Cathartes aura*

SANTA BARBARA ZOO

Western pond turtle, *Actinemys marmorata*

OREGON ZOO

★★★★☆

SO GOOD

Came with super-cute case. Not the fastest, but outlasts every comparable unit. Retractable accessories. Dropped into a pond and still works perfectly!

★★☆☆☆

WHY DID I BELIEVE THEM?!

The seller said the name, giant sea bass, was just "somebody's idea of a funny name for such a small species." After eating everything else in my tank, this brute is now four and a half feet long, must weigh nearly 150 pounds, and it's still growing! The two-star rating is for the seller, not the fish!

Giant sea bass,
Stereolepis gigas

CALIFORNIA SCIENCE CENTER

Spotted garden eel, *Heteroconger hassi*

VANCOUVER AQUARIUM

★★★★☆

UP YOUR GARDEN GAME

Green-thumb approved. These perennials get best results in mild climates. Be warned: slimy residue can be off-putting if you've never grown them before.

★★★★☆

MISLABELED

Warning: this is a straight-up bear. Not a sloth. Not a sloth/bear hybrid. Very good quality, as long as you know what you're getting into.

Sloth bear,
Melursus ursinus

WOODLAND PARK ZOO

King cobra,
Ophiophagus hannah

VIRGINIA ZOO

★★★★☆

Ordered a new "king" cobra, but this female arrived. Does what it's supposed to do otherwise, I guess.

It looks friendly enough at first, but the spines suggest otherwise. I like the departure from the classic fish style and the quill-ity craftsmanship, but be warned: it is prone to random inflating.

Spot-fin porcupinefish,
Diodon hystrix

TEXAS STATE AQUARIUM

Sent away for this product for my grandson, expecting to receive lemon-flavored drink powder used by the NASA astronauts who went to the moon. Instead, it's a yellow fish! DON'T GET FOOLED like me. You can (and should) put it in water, but it does not make the water taste good AT ALL.

Yellow tang,
Zebrasoma flavescens

MARITIME AQUARIUM

Lemur leaf frog,
Agalychnis lemur

MYSTIC AQUARIUM

★★★☆☆

Found upside down, sleeping on the job during the day. Constantly need to remind this guy he runs on fruit flies and crickets, not photosynthesis.

★★★★☆

Calling all stargazers! Stunning! It's a bit fragile and makes a "clicking" sound every once in a while. Other than that, it's quite impressive.

Starry night cracker, *Hamadryas laodamia*

SOPHIA M. SACHS BUTTERFLY HOUSE

Plains zebra, *Equus quagga boehmi*

EL PASO ZOO

PLEASED OVERALL

I needed a large canvas to practice coloring inside the lines. This fit the bill perfectly! Warning: the material doesn't work with colored pencils. Four stars because it would neigh at me and walk away if it didn't like my color scheme.

Zookeepers and educators
Animalis amans

LEHIGH VALLEY ZOO

★★★★★

Very durable. Will work in all weather conditions. Battery power is outstanding. Thrives on free food. This is one that won't let you down, highly recommended.

ACKNOWLEDGMENTS

The Association of Zoos and Aquariums would like to thank the members of the AZA, whose collective creativity made this book possible. The social media staffs at Monterey Bay Aquarium and Oregon Zoo deserve a big thank you, particularly Shervin Hess at Oregon Zoo, who first came up with the idea for #RateASpecies. AZA would also like to thank the all-volunteer AZA Public Relations and Marketing Committees for their assistance in making this book a reality. At AZA, the communications and marketing staff—Ashley Jones, Kristen Corl, Jeff Dow, and Rob Vernon—deserve recognition, as well as AZA President and CEO Dan Ashe for letting his team try new things. And, of course, thank you to the hundreds of animals in this book and the thousands more at AZA-accredited facilities across the world. Whether you fly, swim, walk, or slither, we love telling your stories.

PHOTO CREDITS

Ken Ardell / Toronto Zoo, page 70
Alexis Balinski / Photoshelter, page 143
Rich Barrios, pages 35, 133
Kristina Barroso Burrell, page 39
Amelia Beamish, page 11
Joseph Becker, page 61
Larry Cameron, pages 38, 80
Katie Canady, page 20
Cheyenne Mountain Zoo, pages 43, 85, 98, 111, 144, 150
Claire Cohen, page 139
Shannon Colbert / Tennessee Aquarium, page 126
Jeff Conklin, pages 102, 125, 148
John Cordes, page 82
Dallas World Aquarium, page 91
Dallas Zoo, pages 36, 69, 81
Dennis Dow / Woodland Park Zoo, pages 25, 79, 130
Downtown Aquarium Denver / Landry's, Inc., page 75
Michael Dray / Virginia Zoo, page 156
Edward Durbin Photography, page 129
Jeremy Dwyer-Lindgren / Woodland Park Zoo, page 167
El Paso Zoo, pages 16, 29, 76
Michael Durham, page 31
Michael Durham / Oregon Zoo, pages 12, 37, 53, 68, 83, 136, 164
David Fenn, page 105
Steven Gotz, page 56
Conor Goulding, page 127
Great Plains Zoo & Delbridge Museum of Natural History, pages 55, 157
Greensboro Science Center, page 22
Chris Hartley, page 173
Shervin Hess / Oregon Zoo, page 93
Darris Howe, page 50
Lisa Hubbard, page 32
Michael Jones, page 63
Stephen Knoop, page 116
Chuck Kopczak, PhD, page 165
Sarrah LaSuer, page 134
Roy Lewis, page 87
Lion Country Safari, page 96, 114
Rick LoBello, page 174
Taylor Martin, page 168
Maryland Zoo / Sinclair Boggs, pages 10, 23, 28, 30, 44, 66, 110, 112, 117
Cheryl Miller, pages 137, 172

Nick Moffitt / Blank Park Zoo, page 46
Hailey Moller / Virginia Zoo, page 154
Monterey Bay Aquarium, pages 9, 65, 99, 123
Naples Zoo, pages 17, 106
McKenna Nischke, page 146
Greg Nyquist, pages 14, 40, 47, 109, 121, 135
Oregon Coast Aquarium, pages 158
Amber Paczkowski, page 19
Morgan Powe, page 131
Racine Zoo, pages 18, 72, 90, 95, 142
Frank Ridgley, page 86
Frank Ridgley / Zoo Miami, page 152
Ripley's Aquarium of Myrtle Beach, page 94
Mike Rodgers, page 64
Julie Rohner, page 74
Santa Barbara Zoo / Lauren Gonzales, page 163
Chris Sauder, page 176
Seattle Aquarium, pages 107, 118, 138, 140, 151
Sequoia Park Zoo, page 27
Dave Sigworth / Maritime Aquarium, pages 77, 97, 155, 171
Maria Simmons, pages 34, 119, 161
Mark Simon / Utica Zoo, page 100
Jessica Slater / Peoria Zoo, page 103
Dustin Smith / North Carolina Zoo, page 45
Wayne Smith, page 162
South Carolina Aquarium, page 62
Squam Lakes Natural Science Center, pages 141, 149
Todd Stailey / Tennessee Aquarium, page 153
Kathy Street / Oregon Zoo, page 52
Tennessee Aquarium, page 147
Texas State Aquarium, pages 8, 13, 26, 41, 49, 58, 88, 159, 169
Matthew Thomas Photography, page 71
Adam K. Thompson / Zoo Atlanta, pages 51, 67, 92, 115
Tracy Aviary, page 145
Vancouver Aquarium, pages 73, 166
Edijs Volcjoks / Alamy Stock Photo, page 122
Lauren Wester, page 132
Kat Wille / John Ball Zoo, pages 59, 104
Fort Wayne Children's Zoo, pages 57, 124

The Association of Zoos & Aquariums (AZA) is a 501(c)3 non-profit organization dedicated to the advancement of zoos and aquariums in the areas of conservation, education, science, and recreation. AZA represents more than 230 institutions in the United States and overseas, which collectively draw more than 195 million visitors every year. These facilities meet the highest standards in animal care and provide a fun, safe, and educational family experience. In addition, they dedicate millions of dollars to support scientific research, conservation, and education programs. A percentage of the sales from this volume will go to furthering that mission.